YOUR KNOWLEDGE HAS VALUE

- We will publish your bachelor's and master's thesis, essays and papers

- Your own eBook and book - sold worldwide in all relevant shops

- Earn money with each sale

Upload your text at www.GRIN.com and publish for free

Bibliographic information published by the German National Library:

The German National Library lists this publication in the National Bibliography;
detailed bibliographic data are available on the Internet at http://dnb.dnb.de .

Imprint:

Copyright © 2015 GRIN Verlag, Open Publishing GmbH
Print and binding: Books on Demand GmbH, Norderstedt Germany
ISBN: 9783668331495

This book at GRIN:

http://www.grin.com/en/e-book/342254/water-balance-estimation-of-selected-land-
cover-in-taganibong-watershed

Bryan Allan Talisay

Water Balance Estimation Of Selected Land Cover In Taganibong Watershed, Bukidnon Using GEOWEPP Model

GRIN Publishing

Water Balance Estimation Of Selected Land Cover In Taganibong Watershed, Bukidnon Using GEOWEPP Model

Bryan Allan M. Talisay[1]

[1]Research Associate, Phil-LiDAR 1 Project, College of Forestry and Environmental Science, Central Mindanao University, Musuan, Bukidnon, Philippines.

Abstract

The Geospatial Water Erosion Prediction Project (GeoWEPP v10.2) was tested using data from field survey in three study site within Taganibong Watershed, Bukidnon. This field survey data such as waypoints and soil data was processed and edited through ArcGIS software to prepare for model simulation. Weather information were collected using Automatic Weather Station on an hourly basis at the study site. Climate data was created using Breakpoint Climate Data Generator (BPCDG) which allows direct use of observed storm and other daily standard climate datasets. Model simulation was applied in three land cover to evaluate their abilities in reducing runoff and their effects on other water balance parameters. For the study area, the bamboo area with undisturbed harvesting practices has more effective way in reducing surface runoff based on the results of model simulation. On the other hand, the cultivated area (corn) produce highest surface runoff due to the exposure of soil after harvesting period. Statistical analysis was applied to determine the effects of different land cover to the amount of water balance given the amount of rainfall in monthly basis. Magnitude of rainfall as the primary input to water balance has a huge effects in return to hydrologic processes (percolation, soil water, subsurface lateral flow,etc). This study provides a theoretical basis and technical support for land use zoning, policy implementation, soil and water conservation in upland watersheds.

Keywords: GeoWEPP, Water balance, Taganibong, Watershed

Table of Contents

Background

A watershed is a topographically delineated area that is drained by a river system (i.e., the total land area above some point on a stream or river that drains past that point). It is a hydrologic unit that often is used as a physical-biological unit and a socio-economic-political unit for the planning and management of natural resources (World Forestry Congress, 2003 as cited by Talisay, 2015). As a component of forestry, watershed provides quality and quantity freshwater for unlimited human needs and for the habitat of wild animals. Most of the country's watershed has undergone massive changes over the years and are currently in varied stages of considered degradation thereby needing deliberate rehabilitation efforts. Out of these watershed areas, some are 127 proclaimed watersheds, which are intended to be managed and protected. However, these areas were heavily encroached and subjected to various types of cultivation. Consequently, 90% of these proclaimed watersheds were categorized as hydrologically critical, characterized by degraded biophysical conditions and have become risks to downstream infrastructure (DENR-ERDB, 2010). As evidences of watershed degradation are the physical manifestations of problems such as soil erosion, polluted water runoff, evidence of frequent flooding, sediment-filled channels and reservoirs, and shortages of potable water.

Soil erosion and surface runoff in watersheds have a significant effect on humankind. Thus, it generates serious environmental and economic problems. It is associated with adverse environmental impacts and crop productivity loss (Lal, 1995; Pimentel et al., 1995) that makes its understanding important in assessing food security (Daily et al., 1998) and environmental safety (Matson et al., 1997). It is one of the most serious environmental problems in many areas including upland watersheds in the Philippines.

There have been several models developed to estimate sediment yield, runoff and soil erosion, but in this study, the Water Erosion Prediction Project (WEPP) is being applied. It was developed to estimate soil erosion, water balance and runoff based on specific erosion factors including soil type, climate conditions, ground cover percentage, and topographic condition (Flanagan and Livingston, 1995). The WEPP model calculates sediment yield, runoff, infiltration, and erosion and deposition rates for every day and for multiple time periods (i.e. Monthly or yearly). Since WEPP is process-based model, it requires a great amount of input data to evaluate erosion and sediment yield potentials (Flanagan and Nearing, 1995).

GeoWEPP, a geospatial erosion prediction model, was established to integrate the advanced features of GIS (Geographical Information System) within, such as processing digital data sources and generating digital outputs (Renschler, 2002). In this study, GeoWEPP (v10.2) were used to estimate and predict the water balance of specific study area in Taganibong Watershed.

Methodology

Study Area

The research study was conducted within the three identified land cover in Taganibong Watershed (bamboo, cultivated and fallow), which lies between 124 degrees 57 minutes to 125 degrees 35 minutes east longitude and 7 degrees 48 minutes to 7 degrees 57 minutes north latitudes situated within the Municipality of Maramag and City of Valencia. It covers the Barangays of Guinuroyan, Tugaya, Barobo, Mount Nebo and Lumbo, which is part of Valencia City; Barangay Kisanday and Dologon of the Municipality of Maramag. It has an elevation of 392 meters above sea level (masl). This area has a tropical, maritime climatic condition and belongs to third type having no very pronounced season. Dry season is experienced from November to April while the rest of the year is wet.

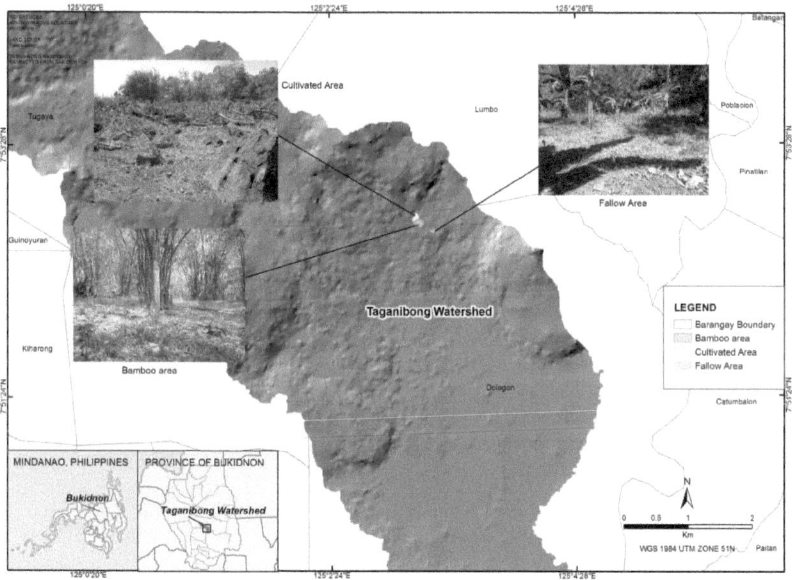

Figure 1. Location Map of the Study Area

Data Collection of Parameter Files

Figure 2 shows the process of WEPP model that needs some parameter data files that were processed and entered to the WEPP model databases. Through GIS, map layers such as DEM, soil and landuse/management map layer were digitized and interpolated using ArcGIS tools. Using the database text files created using a text editor, WEPP and GeoWEPP were linked to run a simulation which has a vital role and would be the basis for interpreting simulation outputs as well as the policy formulation and implementation. Data parameter needed includes climate data, management/land use data, slope data and soil data.

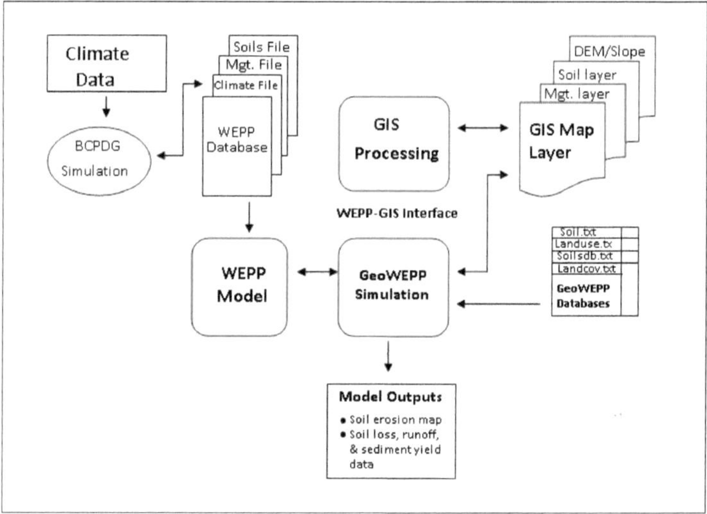

Figure 2. Process flow chart for model simulation, (adopted from Puno, 2014)

Climate Data Collection

Climate data parameters were collected using Automated Weather Station installed in Forest Research and Development Division, Musuan, Bukidnon which includes precipitation data, rainfall intensity, wind speed and direction, air temperature (minimum and maximum), solar radiation, humidity and dew point temperature as well as the date and time. Climate data comprise a one (1) year duration which contained the parameters needed for WEPP simulation. It was consolidated from year 2014 and 2015 due to some insufficiencies of data collected during the year 2014.

Land Cover and Plant Parameters Data Collection

Using a handheld Global Positioning System (GPS), a field survey was conducted to determine the vegetation cover of the land use, area in meters as well as estimate percent rock present in each area. Plant parameters (maximum height, diameter, canopy cover, root-shoot ratio, biomass and etc.) were collected using available data and literatures.

Slope Data Collection

Slope data were obtained using the Digital Elevation Model (DEM) and it was processed in ArcGIS software. Slope information such as segment, the gradient in percent, length, and aspect are the required parameters for slope input file. These parameters are automatically extracted from a DEM in the watershed application using GeoWEPP.

Soil Data Collection

There were three (3) soil sample pits in each landuse collected to represent the soil in the whole area. The depths of the topsoil and subsoil of the soil sample pits were based on the average measurements of the soil files in the associated database available in the WEPP model. In this study, one (1) inch depth of the surface/top soil and two (2) feet for the subsoil. Soil textural classes such as percent sand, silt, and clay, letting in the organic matter content were determined through soil analysis. The soil samples were examined at the City Soil Analysis Laboratory in Malaybalay City, Bukidnon.

Data Preparation and Processing of Database File

There were two application corresponds in WEPP model, the hillslope and the watershed. The hillslope application of the model requires four major input data files; the climate file, the land cover or cropping/management file and slope file, soil file. Moreover, the watershed application of the model needs channel file, and hillslope information file. In GeoWEPP application, the preparation of input files for the WEPP model was automatic, with GIS application software using the graphic feature and computer interface digitization using the computer's mouse. This access is an advantage if the user formulates a DEM, soil and land usage data that represented in GIS map layers.

Digital Elevation Model

Applying the 1,700 waypoints collected over the entire study area using a handheld GPS, the DEM was created and processed using ArcGIS tool. Interpolation of the waypoints was done using interpolation tool Inverse Distance Weight from Spatial Analyst in ArcGIS. The best results from IDW are obtained when sampling is sufficiently dense with regard to the local variation you are attempting to simulate. If the sampling of input points is sparse or uneven, the consequences may not sufficiently represent the desired surface. Thus, large numbers and dense collection of these points were needed to have quality processed raster DEM.

Land cover and Soil Map Layers

The processed DEM was used in creating land cover and soil map layers. The collected waypoints in each perimeter of the land cover served as the basis for land use boundary and it was used for digitizing from one to another point to form a close polygon. Landuse data were entered to attribute table in ArcGIS which describe the existing land cover on each polygon. Using the waypoints collected where the soil sample pits were located, soil map was also created based on the results of the soil analysis. The resulting raster landuse and soil map were converted into the American Standard Code for International Interchange (ASCII) as the final format necessary for the GeoWEPP model simulation.

Climate Database File

Breakpoint Climate Data Generator (BPCDG) was applied based on its advantages over the Climate Generator (CLIGEN) earlier identified by Zeleke et al. (1999). Unlike other climate data generators, BPCDG allows direct use of observed storm and other daily standard climate data sets. With this program, input files required can be created with any text editor. The input can be found in any standard climate station as there is no need for a special climate station, such as the 15-minute precipitation recording stations in the US. Zeleke et al. (1999) as cited by Puno (2009) identified four basic input file names needed to run BPCDG. First is the xxyyyyPL.CSV file containing records of rainfall or storm patterns, about day and month of the year, start and end time of storm event, and the intensity of a storm on a specific day. Second is the xxyyyyCS.CSV containing the date, daily minimum and maximum air temperatures, and wind data, such as strength and direction recorded in 8 and 18 hours. Third is the xxyyyyCL.DAT, which contains Julian day, monthly or annual solar radiation and dew point temperature, and conversion tables for wind data. The fourth file is the xxyyyyST.DAT, which contains the station information such as name, location, elevation, and year of record. In all file names, "xx" is a two-letter code that represents the station name and "yyyy" indicates the year when climate data sets will be compiled.

Landuse/Management Database File

Landuse or crop management file was created and edited using the management editor in WEPP interface which consists of cropland initial condition database, operation database, and plant database. The plant database comprises many parameters such as growth and harvest, temperature and radiation, canopy cover, leaf area and etc. These plant parameters have specific built-in database that can be edited using plant database editor tool.

Soil Database File

Using the Soil Database Editor Tool within the WEPP model, the data on soil textural properties such as organic matter content, percent sand, silt and clay and other soil physical characteristics such as albedo, presence of rock in percent, and layer depth will be entered and stored.

Slope Database File

Created using the DEM in the watershed application using GeoWEPP, the slope information such as segment, the gradient in percent, length, and aspect are the required parameters for slope input file. These parameters were automatically extracted and the resulting output files are saved as individual file within the project subfolder of the GeoWEPP main folder. These files were used when the hillslope option of the WEPP model was applied.

Model Simulation Run

The determination of the temporal deviation of water balance over the different landuses, the hillslope simulations was applied. There were two methods comprises during the simulation such as the offsite or watershed method and the onsite or flowpaths method. The offsite method takes each hillslope in the subwatershed, serves as a representative profile for the hillslope and assigned one soil and one land use to the hillslope. The representative profiles were based on the combined profiles of all the flowpaths found in the hillslope. The flowpaths were then aggregated to create a profile that best represented the hillslope as a whole. For each hillslope, GeoWEPP determined the dominant land use and soil for that hillslope and assigned it to the profile. The WEPP simulation run on each hillslope and the results are compiled. This method is called the offsite assessment because the values reported represent the amount of runoff in each hillslope and evaluated at the outlet point. Report files after the simulation was shown using a text editor and WEPP summarized the output data depending on the options selected before the simulation starts.

Statistical Analysis

Correlation were applied in water balance parameters in each landuses to determine how rainfall affects the increase and decrease of a specified parameter in the study area considering the different land management as well as the existing vegetation within each area. It was also applied to determine the significant variability of the amount of a water balance that occurs among the land management.

Results and Discussion

GIS Layers

After gathering the required data, data preparation and processing were done using ArcGIS tools to provide visualization as well as some interpretations to the datasets which serve as the primary input data for the GeoWEPP model simulation. Shown in Figure 3, the Digital Elevation Model created from collected waypoints and interpolated using Inverse Distance Weight tool in ArcGIS with 2 m by 2 m resolution. The channel network was delineated automatically by GeoWEPP using the Topography Parameterization (TOPAZ) defining the critical source area (CSA) and the minimum source channel length (MSCL) as well as the selection of the watershed outlet for the study area. In this research study, 0.01 hectare for CSA and 20 meters for the MSCL were used. Increasing the value of the MSCL and CSA results to less-detail of the channel network that appear in the DEM.

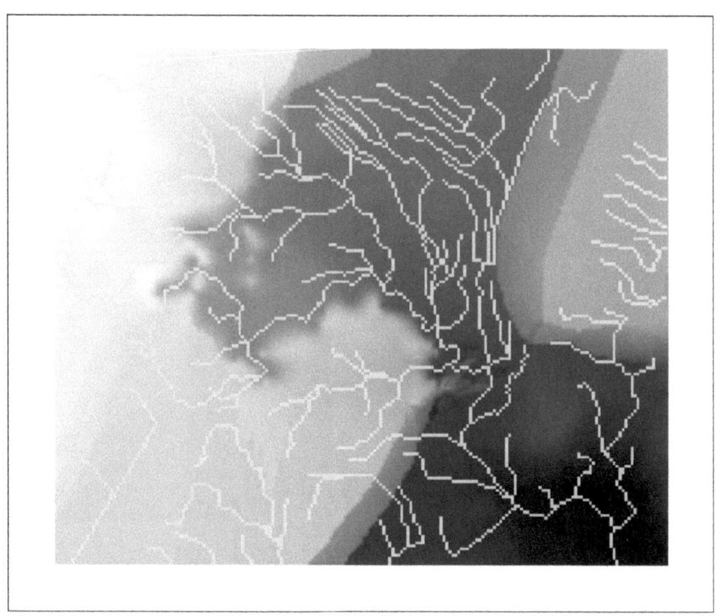

Figure 3. Digital Elevation Model and channel network of the study area

Figure 4 shows the land cover of the study area created from collected perimeter waypoints using Editor tool in ArcGIS to define the perimeter and area. The polygons shown were assigned with cultivated area, bamboo area, fallow area and no data for the outside of the area of interest. The largest area delineated was the bamboo with an area of 0.85 hectare followed by the cultivated with an area of 0.55 hectare and lastly the fallow with an area of 0.11 as the smallest land cover.

Shown in the soil map (Figure 5), there were three (3) soil types assigned in each land cover with identification based on the soil laboratory analysis. The soil properties of the samples were used in creating the soil text file for the GeoWEPP simulation which determines the erodibility of the soil in each land cover. Generally, clay soil type was dominant all over the study area.

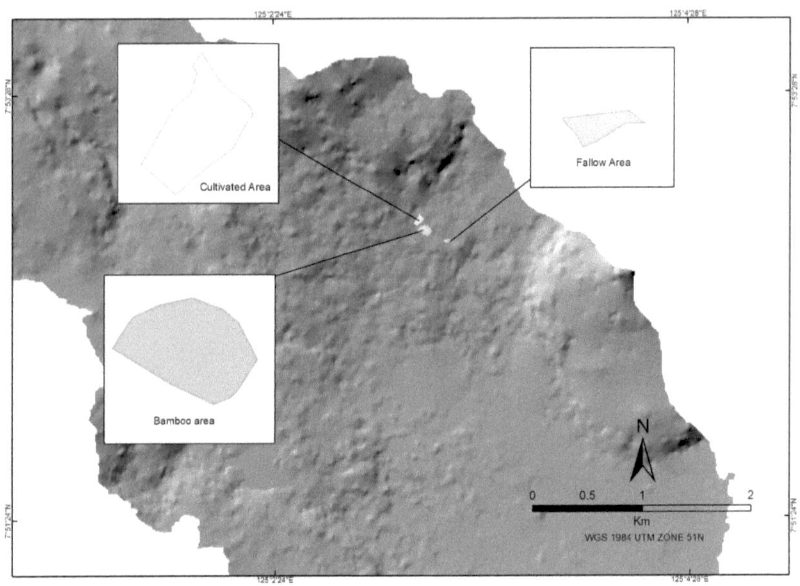

Figure 4. Land cover map of the study area

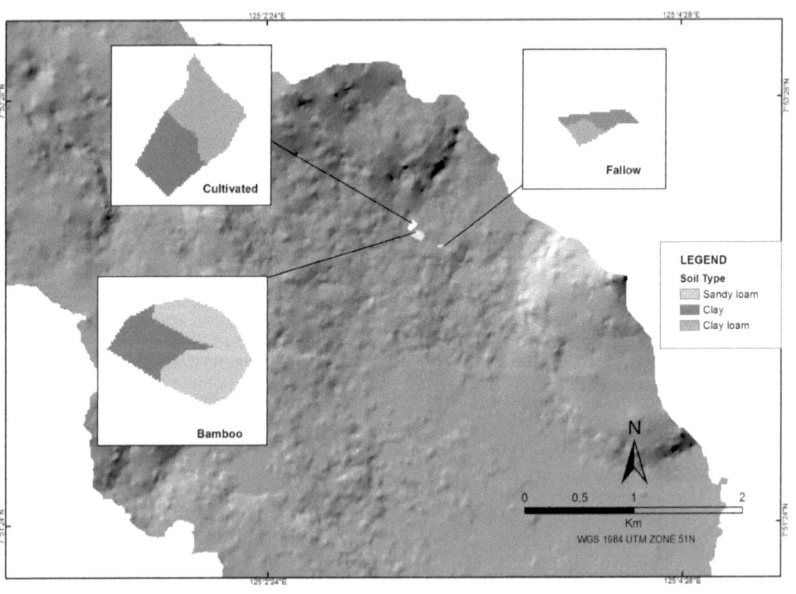

Figure 5. Soil map of the study area

Water balance simulation

The occurrence water balance is affected by several factors, such as climatic conditions (rainfall amount, temperature, pattern, intensity), soil conditions (texture, soil water content, organic matter), land management (vegetation, tillage), and topography (slope, elevation). The changing climate and human activites lead to the discrepancies of the amount of rainfall occur every month. Accuracy of the methods and instruments also affect the quality of the gathered data. The results of the simulation covers the temporal and spatial distribution of predicted monthly water balance within the study area.

In table 1 shows the predicted monthly rainfall which varies throughout the year. For one year period of simulation, monthly total rainfall ranges from 79.1 mm to 304.6 mm with much precipitation (75% of the annual rainfall) occurring between May and November with generated annual rainfall amount of 2122.7 mm.

Generally, high amounts of predicted runoff occurring between the months of May to November. However, the month with the highest surface runoff, i.e., October, did not corresponds to the month with the highest monthly rainfall, except for the cultivated area. This agrees with the findings of Pieri et al. (2007) in Appennines mountain range in Italy. Highest annual predicted runoff occurred in cultivated area followed with the bamboo and lastly, the fallow area.

Table 1 also shows the predicted monthly runoff of the bamboo area ranging from 0.3 to 71.5 mm and has a total annual runoff of 211.0 mm. Predicted runoff in cultivated area ranging from 5.0 to 157.1 mm and has the annual runoff of 985.7 mm. In addition, an annual predicted runoff of 474.7 mm occurred in fallow area which ranges from 1.5 to 97.3 mm. This indicates that bamboo area is more effective in reducing runoff as compared with other vegetation such as cultivated land or grasslands. Bowyer et al. (2014) reported that bamboo stands stabilizes erodible slopes and flood-prone watersheds. It stabilizes the earth with its erosion-preventing root and enhances the health and purity of soils (Rivera, 2000). Types of vegetation has an effective way in reducing the surface runoff and agreed with the of Zhang et al. (2015) that a 5-year-old mixed grass-shrub-arbor forest is the most effective in preventing runoff in a slope plot area than the legume plants in Beijing, China.

Extreme soil evaporation was occurred in cultivated area with a value of 598.2 mm, annually then followed with Fallow area having a annual soil evaporation of 457.4 mm. Lastly, the bamboo area with a value of 325.5 mm. Vegetation pattern can exhibit a spatio-temporal variation of the soil evaporation. The presence of canopy cover and litter control the rate of evaporation in the soil as reported by Villegas et al. (2009) which agrees to the result of the study. Another study reported by Ansari (2003) states that the loss of moisture through evaporation from the soil in an expose area (agricultural land) is approximately five times as greater as loss of moisture in forest covered area.

Table 1. Predicted water balance within the study area

Land Cover	Month	Rainfall mm	Runoff mm	Deep Percolation mm	Soil Evaporation mm	Subsurface Lateral flow mm	Soil Water mm
Bamboo	Jan	79.1	5.0	0.0	20.1	0.0	3.1
	Feb	97.8	4.0	0.0	22.5	0.0	6.6
	Mar	110.3	11.0	0.3	29.3	0.0	12.6
	Apr	134.8	1.1	31.8	31.6	0.0	16.1
	May	304.6	22.6	128.9	39.1	1.2	19.8
	Jun	174.6	9.4	101.1	23.8	0.09	18.9
	Jul	242.8	39.1	120.1	30.3	1.5	19.4
	Aug	150.4	0.3	95.2	19.5	0.34	18.2
	Sep	238.8	30.6	126.6	16.8	1.77	19.9
	Oct	229.8	71.5	105.0	31.9	0.51	18.6
	Nov	260.3	14.4	117.8	35.3	0.32	19.4
	Dec	99.4	2.0	63.8	27.9	0.03	17.3
	Total	**2122.7**	**211.0**	**890.7**	**328.5**	**5.76**	**189.8**
Cultivated	Jan	79.1	5.0	0.0	38.5	0.0	1.2
	Feb	97.8	14.9	0.0	41.3	0.0	2.8
	Mar	110.3	26.6	0.06	53.6	0.0	5.6
	Apr	134.8	45.2	6.7	56.0	0.0	6.9
	May	304.6	157.1	21.9	71.5	1.1	8.9
	Jun	174.6	84.9	16.2	42.6	0.1	8.5
	Jul	242.8	137.3	18.5	54.7	0.5	8.7
	Aug	150.4	70.6	15.5	37.1	0.3	7.7
	Sep	238.8	138.9	19.7	30.2	0.7	8.9
	Oct	229.8	145.9	17.3	58.7	0.5	7.8
	Nov	260.3	124.6	18.8	63.2	0.3	8.7
	Dec	99.4	34.7	11.5	50.7	0.0	7.3
	Total	**2122.7**	**985.7**	**146.6**	**598.2**	**3.4**	**82.8**

Land Cover	Month	Rainfall mm	Runoff mm	Deep Percolation mm	Soil Evaporation mm	Subsurface Lateral flow mm	Soil Water mm
Fallow	Jan	79.1	1.5	0.0	29.7	0.0	2.7
	Feb	97.8	6.0	0.3	31.5	0.0	6.2
	Mar	110.3	11.3	0.0	40.9	0.0	12.2
	Apr	134.8	18.8	13.2	42.8	0.0	15.8
	May	304.6	70.2	37.4	54.6	1.9	14.2
	Jun	174.6	35.3	25.7	32.5	0.1	18.6
	Jul	242.8	72.1	30.2	41.7	0.8	19.0
	Aug	150.4	25.6	23.2	28.3	0.5	12.6
	Sep	238.8	69.8	33.7	23.1	0.2	19.6
	Oct	229.8	97.3	26.5	44.8	0.8	13.1
	Nov	260.3	56.1	31.2	48.3	0.5	14.0
	Dec	99.4	10.6	15.2	38.7	0.0	17.2
	Total	**2122.7**	**474.7**	**236.9**	**457.4**	**4.8**	**165.3**

The predicted annual subsurface lateral flow has a very small amount occurred in each land cover. The highest occurred in bamboo with a value of 5.76 mm followed by fallow area with value of 4.8 mm and lastly the cultivated area with a value of 3.4 mm. WEPP-modelled average annual subsurface lateral flow was below 6 mm over the three land covers, which were less than 1% of the total rainfall. Highest percolation rate occurred in bamboo area with an annual of 890.7 mm followed with fallow area with annual percolation of 236.9 mm while the fallow area has the lowest annual percolation of 146.6 mm. Simulated annual deep percolation varied from 146.6 to 890.7 mm over the land covers which account for 7-41% of the total rainfall and in most cases exceeding the surface runoff.

The insignificant subsurface lateral flow was due to not setting restrictive layer with low permeability in the model. As observed, the result shown a general agreement with the report of Pieri et al. (2007) which states that setting a restrictive layer in the with less impermeability causing the deep percolation decreases and the subsurface lateral flow increases in annual basis.

Changes of soil water greatly affect tree species diversity and forest canopy structure. It is also one of the most important factors controlling hydrological processes (Castillo et al., 2003 as cited by Wang et al., 2013). The predicted amount of soil water in bamboo area has the highest value of 189.8 mm in annual basis while the fallow area has the value of

165.3 mm and lastly the cultivated area as the lowest soil water value of 82.8 mm. The result was in agreement with Luscher and Zurcher (2003) states that the soils under natural forests tend to be relatively porous because trees loosen the soil and accumulate more organic matter with high infiltration rates. Thus, forest cover influences the water retention capacity in forest sites, and in turn increases the overall water storage capacity.

Statistical Analysis

This study determine the effects of different land cover to the amount of water balance given the amount of rainfall in monthly basis. The predicted water balance parameters was correlated using the rainfall data to determine the significant relationship within different study area shown in Table 2.

Table 2. Correlation analysis of water balance of the different land cover

Pearson Correlation		Runoff	Deep Percolation	Soil Evaporation	Subsurface Lateral Flow	Soil Water
Bamboo	**Rainfall**	0.609*	0.883**	0.510	0.755**	0.742**
	Sig. (2-tailed)	0.037	0.000	0.90	0.005	0.006
	N	12	12	12	12	12
Cultivated	**Rainfall**	0.973**	0.880**	0.499	0.886**	0.762**
	Sig. (2-tailed)	0.000	0.000	0.099	0.000	0.004
	N	12	12	12	12	12
Fallow	**Rainfall**	0.901**	0.915**	0.497	0.803**	0.519
	Sig. (2-tailed)	0.000	0.000	0.100	0.002	0.084
	N	12	12	12	12	12

*Correlation is significant at the 0.05 level (2-tailed).

**Correlation is is significant at the 0.01 level (2-tailed).

In Table 2, there is a very strong positive relationship between the rainfall and runoff within cultivated and fallow which is highly significant at alpha level 0.01. On the other hand, there is a strong relationship between rainfall and runoff in bamboo area which is significant at alpha level 0.05. This indicates that there is direct propotion of rainfall and runoff, as the runoff increases the rainfall also increases and otherwise.

Moreover, there is a very strong positive relationship between the rainfall, deep percolation and subsurface lateral flow within different land cover. It is highly significant at alpha level 0.01. This means that when the magnitude of rainfall increases, the deep percolation and subsurface lateral flow also increases.

There is a moderate positive relationship between the rainfall and the soil evaporation within the study area which is not sigficant in alpha level 0.01 and 0.05. This implies that when the rainfall either increases or decreases, the pattern of soil evaporation will be variable in its magnitude. The occurrence of soil evaporation has factors to be considered like the temperature, the land cover and the rainfall amount which affect in its amount.

There is a strong positive relationship between the rainfall and soil water for the bamboo and cultivated area which is highly significant at alpha level 0.01. On the other hand, there is a strong positive relationship between the rainfall and soil water in fallow area which is not significant at alpha level 0.01 and 0.05. This denotes that when the magnitude of rainfall increases, the magnitude of soil water also increases for the bamboo and cultivated while soil water in the fallow area is variable in a given amount of rainfall.

Conclusion

Due to the large amount of parameter input files needed by WEPP, relevant secondary data were gathered and entered to the model which still results to reliable prediction values though it is still need to be validated particularly the water balance ouputs. The GeoWEPP model has the ability to quantify the temporal and spatial distribution of water balance within study area considering the complex prevailing condition.

The GeoWEPP model, which incorporates digital elevation data with the WEPP version 2012.8 to generate water balance outputs for complex land management including climatic, soil and topographic features, has the potential to be a useful management tool for forestry applications specifically in watershed management. Therefore, watershed managers can locate the problematic areas in a watershed and implement necessary precaution measures to minimize severe runoff and to determine other water balance parameters.

The result indicates that the bamboo area has more effective way in reducing runoff than the other land cover. Over the long term, vegetation types has significant effects in increasing soil organic matter, improve soil physical properties and soil anti-erodibility, and reduce runoff and erosion to a tolerable level. This study provides a theoretical basis and technical support for land use zoning, policy implementation, soil and water conservation in upland watersheds.

Literature Cited

A. Ansari, "Influence of forests on environment", XII World Forestry Congress, Quebec City, Canada, 2003.

B. A. Talisay, "Runoff and water balance simulation in three different land uses within Taganibong watershed using GeoWEPP model", College of Forestry and Environmental Science, Central Mindanao University (Undergraduate Thesis), 2015.

Bowyer et. al, "Bamboo Products and their environmental impacts", NSW Department of Land and Water Conservation, Sydney, Australia, 2014.

C. Renschler, *"Geo-spatial Interface for the Water Erosion Prediction Project GeoWEPP ArcX"*, 2002. Available: http://www.geog.buffalo.edu/~rensch/geowepp/ GeoWEPP%20Manual_files/frame.htm. Accessed March, 2016.

C. Wang, C. Zhao, Z. Xu, Y. Wang, H. Peng, "Effects of vegetation on soul water retention and storage in semi-arid alpine forest catchment", *Journal of Arid Land*, 5, 207-219, 2013.

D. C. FLANAGAN and M.A. NEARING, NSERL Report No. 10, *"WEPP user summary: USDA-Water Erosion Prediction Project (WEPP)"*, USDA-ARS National Soil Erosion Research Laboratory, 1995.

D. C. Flanagan and S.J. Livingston, NSERL Report No. 11, *"WEPP user summary: USDA-Water Erosion Prediction Project (WEPP)"*, USDA-ARS National Soil Erosion Research Laboratory, 1995.

D. Pimentel, et. al., "Environmental and economic costs of soil erosion and conservation benefits", *Science* 267, pp. 1117–1123, 1995.

DENR-ERDB, Report : "Compendium of Rehabilitation Strategies for Damaged, Critical & Marginal Watershed Areas", Ecosystems Research and Development Bureau, Department of Environment and Natural Resources, College, Laguna 4031, 2010.

G. Zeleke, T. Winter And D.C. Flanagan, BPCDG: BCDG for WEPP using observed standard weather data sets. USDA-ARS National Soil Erosion Research Laboratory, West Lafayette, Indiana, 1999. Available online: http://topsoil.nserl.purdue.edu/weppwin, (accessed on March 2014).

G. R. Puno, "Runoff and sediment yield modeling using GeoWEPP in Mapawa Catchment", *CMU Journal of Science,* Musuan, Bukidnon. Vol. 18, pp. 49-70, 2014.

G. Daily, et. al., "Food production, population growth, and the environment", *Science* 281, 1291–1292, 1998.

G.R. PUNO, *"Application of GeoWEPP Model in Simulating Runoff and Soil Loss of Mapawa Watershed in Bukidnon, Philippines"*, Unpublished Dissertation, University of the Philippines, Los Baños, Laguna, 2009.

J. C. Villegas, D. Breshears, C. Zou and D. Law, "Ecohydrological controls of soil evaporation in deciduous drylands: How the hierarchical effects of litter, patch and vegetation mosaic cover interact with phenology and season", *Journal of Arid Environments,* 74, 595-602, 2010.

L. Pieri , M. Bittelli , J. Wu , S. Dun , D. Flanagan , P. R. Pisa , F. Ventura , F. Salvatorelli, "Using the Water Erosion Prediction Project (WEPP) model to simulate field-observed runoff and erosion in the Apennines mountain range, Italy", *Journal of Hydrology*, 336, 84– 97, 2007.

L. Zhang, J. Wang, Z. Bai and C. Lv, "Effects of vegetation on runoff and soil erosion on reclaimed land in an opencast coal-mine dump in loess area", *CATENA*, 128, 44-53, 2015.

M. Rivera, "Philippine National Report on Bamboo and Rattan", ERDB-DENR, 2000. http://www.inbar.int/documents/country520report/philippine.htm(Accessed on November 15, 2014)

P. A. Matson, W. J. Parton, A. G. Power, & M. J. Swift, "Agricultural intensification and ecosystem properties", *Science* 277, 504–509, 1997.

P, Luscher and K. Zurcher. 2003. Flood Protection in Forests. Report of the Bavarian State Institute of Forestry, Report No. 40. Freising: Bavarian State Institute of Forestry.

R. Lal, "Erosion-crop productivity relationships for soils of Africa". *Journal of Soil Science Soc. Am.* 59, 661– 667, 1995.

V. M. Castillo, A. Gomez-Plaza, M. Martínez-Mena, "The role of antecedent soil water content in the runoff response of semiarid catchments: a simulation approach", *Journal of Hydrology*, 284: 114–130, 2003.

World Forestry Congress, Report: *Forestry Projects Aimed at Watershed   management*, FAO, Forestry Department, Rome, Italy, 2003.